AF228303

Searchlight BOOKS™

Hunting and Fishing

Ice Fishing

Diane Bailey

Lerner Publications • Minneapolis

Lerner Publications Company
An imprint of Lerner Publishing Group, Inc.
241 First Avenue North
Minneapolis, MN 55401 USA

For reading levels and more information, look up this title at www.lernerbooks.com.

Main body text set in Adrianna Regular.
Typeface provided by Chank.

Library of Congress Cataloging-in-Publication Data

Names: Bailey, Diane, 1966– author.
Title: Ice fishing / Diane Bailey.
Description: Minneapolis : Lerner Publications, [2024] | Series: Searchlight books - hunting and fishing | Includes bibliographical references and index. | Audience: Ages 8–11 | Audience: Grades 4–6 | Summary: "Did you know people fish even during the cold days of winter? In the winter, it is known as ice fishing. From the history of ice fishing to fishing basics, learn all about the sport"— Provided by publisher.
Identifiers: LCCN 2022043943 (print) | LCCN 2022043944 (ebook) | ISBN 9781728491585 (library binding) | ISBN 9798765603772 (paperback) | ISBN 9798765600511 (ebook)
Subjects: LCSH: Ice fishing—Juvenile literature.
Classification: LCC SH455.45 .B35 2024 (print) | LCC SH455.45 (ebook) | DDC 799.12/2—dc23/eng/20220928

LC record available at https://lccn.loc.gov/2022043943
LC ebook record available at https://lccn.loc.gov/2022043944

Manufactured in the United States of America
2-1010438-51107-12/14/2023

Table of Contents

Chapter 1

FISHING ON ICE

Out on the lake, a fisher is looking for his next catch. He does not need a boat because it is the middle of winter, and the lake is frozen over with ice. Instead, the fisher walks carefully between several holes he has drilled into the ice. At each one, he has dropped a baited fishing line into the water. Suddenly, a bright red flag pops up at one of the holes. That means a bite! The fisher hurries over and begins to pull in the line. Dinner is on the hook.

Ice fishing is a lot different from fishing in the
summer, but many anglers like the idea of being able to
fish year-round. They just need a few new skills and a
lot of patience.

ICE FISHING KEEPS FISHING FUN ALL WINTER LONG.

A lot of the equipment used in ice fishing is the same used for regular fishing. Fishers need rods, reels, and jigs, or lures. They also need an auger to drill holes in the ice, a measuring tape to determine how thick the ice is, and a scoop to clear away broken ice and slush. Some people use a sonar device to help find fish.

It's Cold Out and the Fishing Is Good

Some types of fish don't like the cold and become sluggish in winter. That makes them harder to catch. But some species, like yellow perch, trout, bass, and crappie stay active. They are favorites among ice fishers. Larger fish such as pike and walleye are also good targets. Most ice fishing is done in freshwater ponds and lakes, which freeze faster than saltwater. But it is possible to ice fish in some shallow saltwater areas.

ICE FISHERS CATCH MORE FISH WHEN THEY DRILL MORE HOLES.

Summer fishers can easily move around on the water, looking for the best places to fish. Ice fishers must pick their spots right at the start. But they can drill many different holes around the ice. That lets them fish in several places at once.

Anglers who spend a lot of time on the ice often set up a shelter that is similar to a camping tent. This protects them from cold and snow. Some shelters are attached to a base that is a sled. That makes it easier for fishers to drag their stuff around when they are looking for fish.

Chapter 2

STAYING SAFE ON THE ICE

The first job for any ice fisher is to make sure the ice is thick enough to walk on safely. Some parts of a lake will freeze faster than other parts, and some parts will have thicker ice than others. For example, shallow water at the shoreline freezes before deeper water in the middle.

Water with a current in it also takes longer to freeze. Snow that is on top of ice acts like a blanket, keeping it warm. That means ice that is under snow is usually weaker.

ICE FISHERS MUST BE CAUTIOUS WHEN VENTURING OUT ON THE ICE.

Testing the Ice

Ice fishers begin testing the ice a few feet from the shore, where the water is shallow. That way, if they break through the ice, they can get out easily. First, they step lightly on the ice. If it holds up, they jump and stomp on it. If it is still hard, they move out a few more feet and do it again. When they are sure the ice is solid, they walk out a little farther and drill a test hole with their auger.

Ice that has melted and refrozen is not as strong.

For one person, the ice should be at least four inches (10.2 cm) thick. To be safe, ice should be a bit thicker for groups of people. Ice fishers using a snowmobile for traveling on the ice need to have about eight inches (20.3 cm) of ice. Cars or small trucks need even more.

There is always a risk that ice is weaker than it seems. Ice fishers wear personal flotation devices (PFDs) such as a life jacket. These keep them afloat if they break through the surface. Some ice fishers even wear full-body suits that float and are made of special material that will keep them warm if they fall in. Many ice anglers also carry ice picks. They can stick them into the ice and pull themselves out in an emergency.

A person uses an ice pick to pull themselves out of the water.

STEM Spotlight

Most water has tiny air bubbles in it. When the water freezes, that air gets trapped inside and makes it look cloudy. Air also makes ice weaker. Ice is formed with crystals. When there is a lot of air in the ice, there is less room for the crystals to hold it together. When water freezes slowly, there is more time for air to settle out of the ice. That makes slow-forming ice clearer and stronger.

THE BASICS OF ICE FISHING

Anglers sometimes call ice fishing hardwater fishing since ice is water that has become solid. But fish don't live inside ice, of course. They live in the water underneath the ice. Getting through the ice is the first step.

In ancient times, people chopped holes in the ice with an axe or chisel. Many fishers find that augers are better tools. Augers look like oversized corkscrews. They have spiral blades that drill through the ice. Some are operated by hand, but there are also electric or gas-powered ones.

An experienced ice
fisher can use an auger
to drill through ice in
less than a minute.

The size of the hole depends on the size of the fish an angler is after. An eight-inch (20.3 cm) hole is big enough for most fish—and too small to accidentally fall into.

Finding Fish

Every fishing trip starts with getting information. To find fish, many anglers use a sonar device. Sonar stands for sound navigation ranging. Sonar devices send sound waves through the water. These waves bounce off objects in the water—like fish or underwater reefs—and come back. Where the sound waves traveled, and how long they took, tells anglers where to look for fish.

Using advanced fish finding technology helps fishers plan where to fish.

Fish usually hang out around rocks, reefs, or sandbars, where there is more food. In winter, they spend more time near the bottom of ponds and lakes, because the water is a little warmer. Fishers decide how deep to fish and then measure the depth of the water by sounding. They drop a small weight attached to a line and let it sink to the bottom. The amount of line used up shows them how deep the water is.

Getting a Bite

Most ice fishers use tip-ups. A tip-up is a simple frame that sits on the ice. It has a trip wire with a small flag attached. Anglers tie their fishing reel to this wire. When a fish bites, it pulls the line. That sets off the trigger and releases the flag. That way, the fisher doesn't have to wait by the hole. The flag tells him when there's some action.

If things are slow, anglers may also try to get the fish's attention by jigging. The jig is the lure on the end of the line. Jigging, or bobbing the bait up and down, often attracts fish.

Fishing History

Ancient ice fishers used spears. They lay on their stomachs over their hole and dropped in a lure carved from wood or a shell. They jigged the lure to get the attention of fish. When one swam by, ice fishers speared it. But sunlight that reflects off ice makes it difficult to see into the water. Fishers built huts around their holes to block out light. Huts let in just enough light to see. Darkhouse spearfishing is popular with some anglers today.

Spearfishing was popular among Native American nations in northern climates.

Chapter 4

CONSERVATION

In many places, fish habitats are in trouble. They are being damaged by pollution. Climate change is making some water too warm for fish to live. And when people fish too much in a certain area, it can threaten the whole population. But there are ways to help.

Anglers buy licenses that give them permission to fish in public waters. Youth often don't have to get a license. State agencies use that money to help protect fish and the environment. They clean up and improve habitats, and study fish to see if they are healthy. They also raise fish to stock lakes.

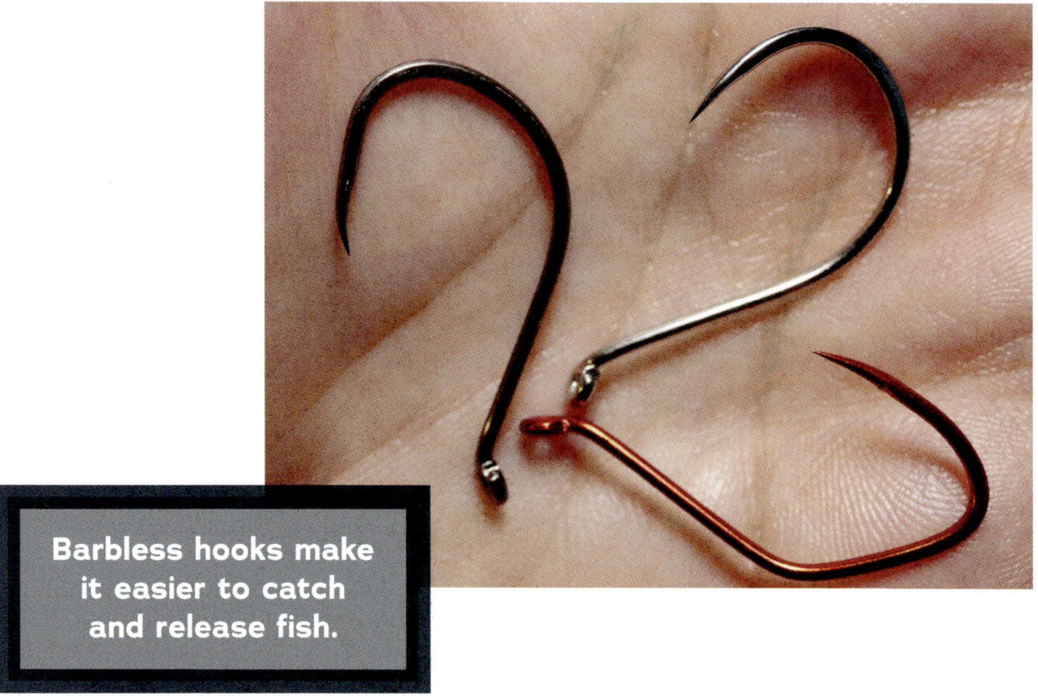

Keeping the water clean is the first step to keeping fish healthy. It is a good practice to not use fishing hooks or lures made from lead. Lead is a poisonous metal. It makes fish—and people—sick. A fish that bites a lead hook gets that poison in its body. It can then be transferred to a person who eats the fish.

Pick and Choose

Fishing regulations are not as strict as hunting land animals. But most places have rules about what types of fish are allowed to be taken. There are also limits about how many fish can be harvested in a day, and how large they can be. Any endangered or protected species are off-limits. Anglers also know to throw back any especially big fish. Larger fish produce more offspring and are important for keeping the whole population alive. Many anglers throw back any fish, even small ones, if they don't plan to eat them. This is called catch and release fishing. It's a way to enjoy ice fishing without doing long-term damage.

No one wants to fish in polluted water or catch sick fish. Good conservation helps keep the environment healthy and makes fishing more enjoyable. And it helps make sure there will be enough fish for everyone!

Fishing Tips

- Avoid ice that looks black or gray. Ice that looks blue is thicker and stronger.

- Change the depth of your line occasionally to find where fish are biting the most.

- Regularly use a scoop or skimmer to clear new ice and keep your hole clear.

- Focus on catching one type of fish at a time. Use a line and bait that fits that particular fish.

Glossary

angler: another name for a fisher

auger: a kind of drill with a corkscrew shape

catch and release: a method for fishing that does not kill fish

conservation: protecting resources like land and water and using them wisely

habitat: the environment a particular animal lives in

hardwater fishing: a nickname for ice fishing

jig: a lure; also the act of bobbing a lure up and down

population: the total number of a type of animal

sonar: a system that uses sound waves to determine the location of other objects

sounding: measuring the depth of the water

tip-up: a frame used to hold a fishing reel on the ice

Learn More

Britannica Kids: Fishing
https://kids.britannica.com/students/article/fishing/274328

Doyle, Abby Badach. *Ice Fishing*. New York: Gareth Stevens, 2023.

How Stuff Works: Ice Fishing
https://adventure.howstuffworks.com/outdoor-activities/fishing/
fish-conservation/responsible-fishing/ice-fishing-adventure-fishing.
htm?utm_source=howstuffworks&utm_medium=recirc

Kiddle: Ice Fishing Facts for Kids
https://kids.kiddle.co/Ice_fishing

Mazzarella, Kerri. *Ice Fishing*. New York: Crabtree Publishing, 2023.

Reeves, Diane Lindsey. *Freshwater Fishing*. Minneapolis: Lerner
Publications, 2024.

Index

Photo Acknowledgments

Image Credits: p. 5; Stephen McSweeny/Dreamstime, p. 6; Dmitry Sytnik/Dreamstime, p.7; Nathan Allred/Dreamstime, p. 8; Vladimir Salman/Shutterstock, p. 9; Onepony/Dreamstime, p. 11; Vitalliy/Shutterstock, p. 12; Dmitry Markov/Dreamstime, p. 13; Yulan/Dreamstime, p. 14; Valerie Loiseleux/iStock Photos, p. 15; Stone36/Shutterstock, p. 17; GROGL/iStock Photos, p. 18; wanderluster/iStock Photos, p. 19; Yauheni Labanau/Dreamstime, p. 20; ToKa74/Shutterstock, p. 21; Steve Callahan/Dreamstime, p. 22; Kalvis Kalsers/Dreamstime, p. 23; bauhaus1000/iStock Photos, p. 25; Debspoons/Dreamstime, p. 26; Hailshadow/iStock Photos, p. 27; Truecash2k8/Dreamstime, p. 28; Varnak/Shutterstock, p. 29; Amelia Martin/Dreamstime.

Cover: Anna Kosolapova/Dreamstime.